TEXTURE

Rebecca Kraft Rector

Enslow Publishing
101 W. 23rd Street
Suite 240
New York, NY 10011
USA
enslow.com

Words to Know

atom A tiny bit of matter.

chemical Having to do with chemistry.

chemistry The science that deals with the properties of matter and how it forms and changes.

gas A kind of matter that has no permanent shape, like air.

liquid A kind of matter that can move freely, like water.

physical Having to do with being able to be touched or seen.

properties The qualities or features of something.

solid A kind of matter that is firm and keeps its shape.

Contents

WORDS TO KNOW 2
WHAT IS MATTER? 5
COMMON FORMS OF MATTER 7
PROPERTIES OF MATTER 9
THE PROPERTY OF TEXTURE 11
TOUCH IT! 13
SOLID TEXTURE 15
LIQUID TEXTURE 17
DIFFERENT BUT THE SAME 19
MORE THAN ONE 21
ACTIVITY: DOUBLE TROUBLE! 22
LEARN MORE 24
INDEX 24

Look around the room. Look out the window. Matter is everywhere!

What Is Matter?

Matter is everything around you. All things are made of matter. Tiny bits of matter are called atoms. Atoms join together to make molecules.

Fast Fact

Even people are made of matter.

Can you spot the different kinds of matter? The train is a solid. The water is liquid. The steam from the train engine is gas.

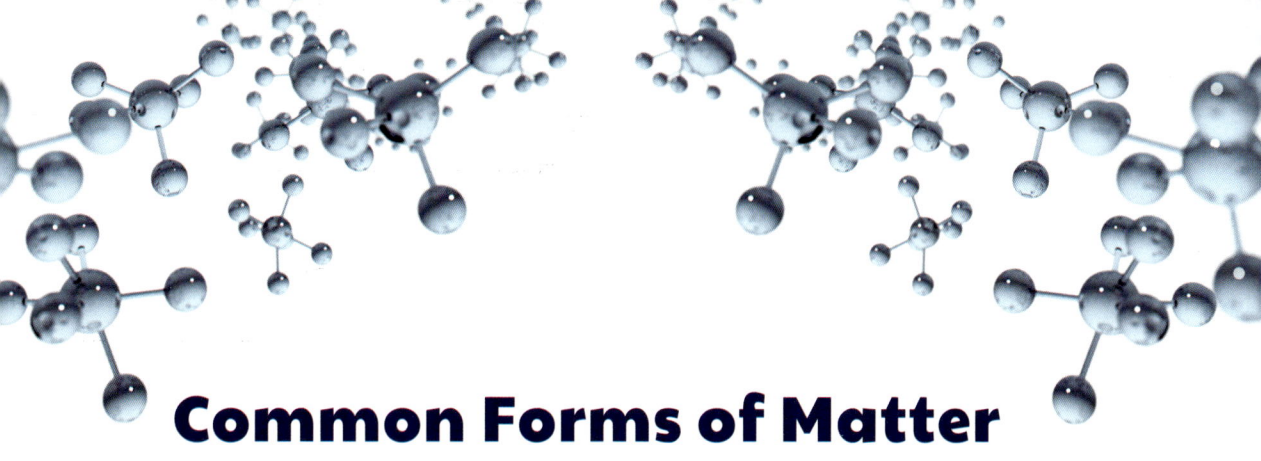

Common Forms of Matter

Matter has different forms. Matter can be solid. A flower is a solid. Matter can be liquid. Tea is a liquid. Matter can be a gas. Steam is a gas.

Fast Fact
Atoms are packed tightly together in a solid.

Look out for poison ivy! The plant can make you itch because of its chemical properties.

Properties of Matter

Properties tell about matter. Physical properties tell how it acts, looks, and feels. Is it big or small? Hard or soft? Chemical properties let matter change. An example is how easily something can become a poison.

A coral reef in the ocean has lots of textures. The coral can be very bumpy.

The Property of Texture

Texture is a physical property. It tells what something feels like. It tells about the outside of things. All solids have texture. Liquids have texture, too.

Have you ever stepped barefoot in the mud? It can have a very slimy texture.

Touch It!

You discover texture through touch. Touch something with your fingers. Touch the ground with your feet. Touch food with your tongue. How does it feel? Sticky? Rough? Mushy?

Fast Fact

Your skin sends messages to the brain about texture.

This fuzzy squirrel is enjoying both rough and smooth nuts.

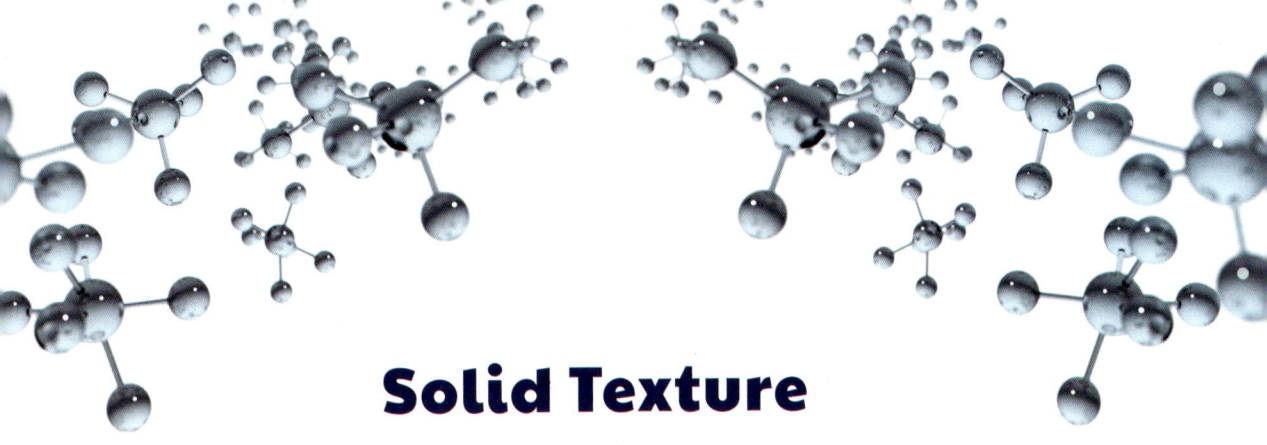

Solid Texture

Solids always have texture. Glass is smooth. An alligator is bumpy. Sand is gritty. Teddy bears are fuzzy. Bricks are rough. Cotton balls are soft.

FAST FACT
You can sort by texture.

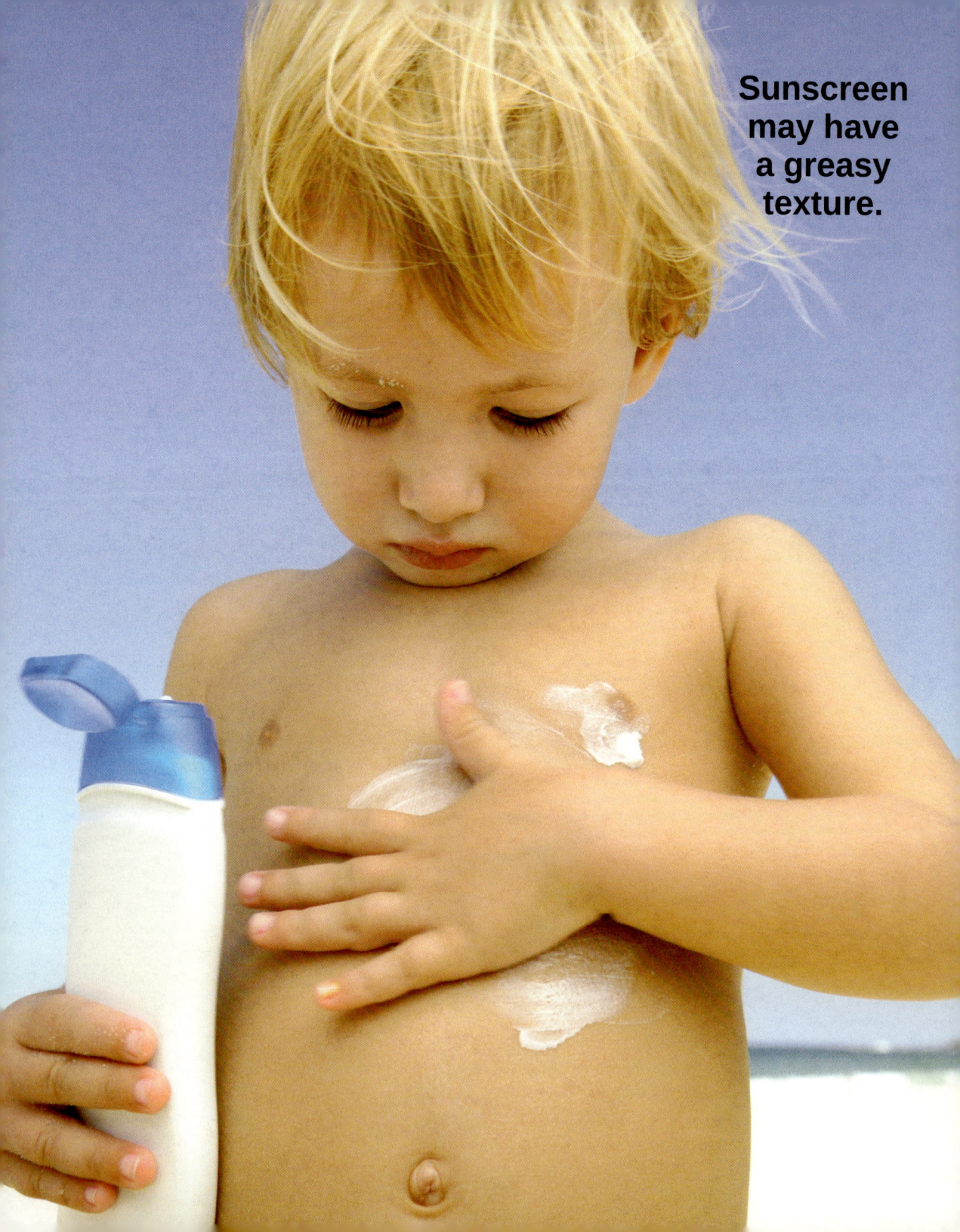
Sunscreen may have a greasy texture.

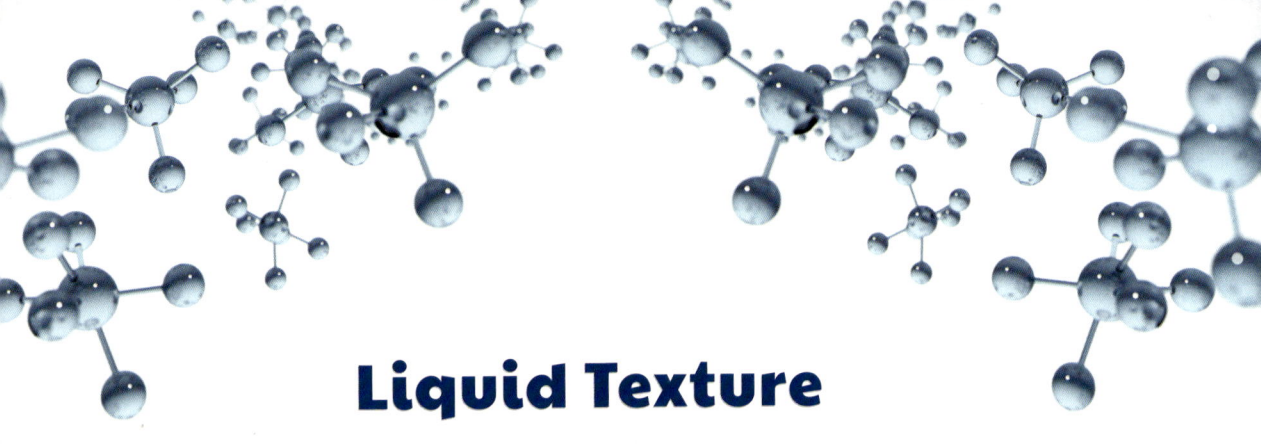

Liquid Texture

Liquids have texture. They can be slippery like oil. They can be gooey like honey. They can be sticky like syrup. They can be greasy like lotions.

Fast Fact

Slippery liquids tend to flow quickly.

**The ice and the skates are smooth.
The girl's coat and hat are soft.**

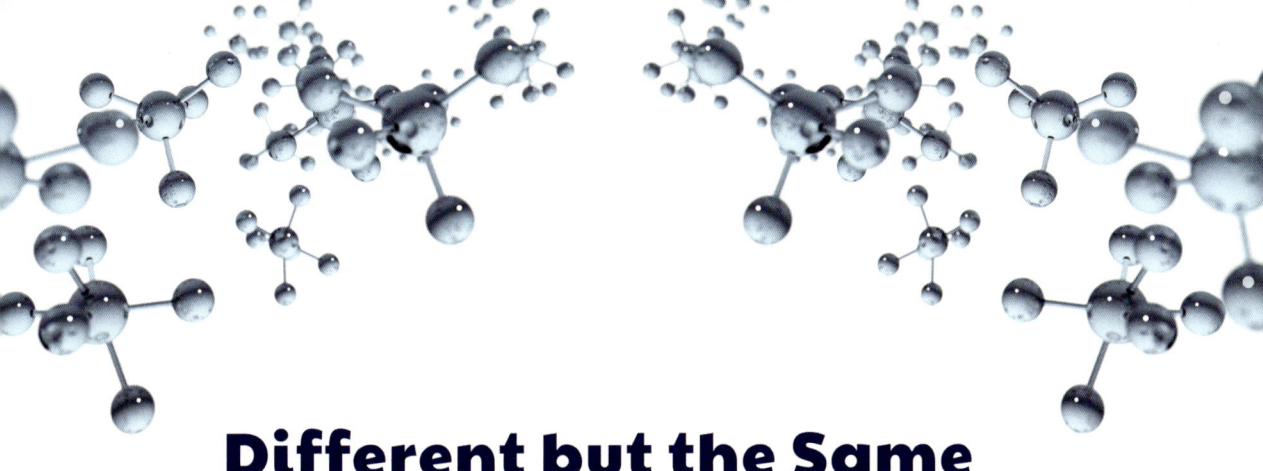

Different but the Same

Different things can have the same texture. A cactus can be prickly. So can a hairbrush. Rocks and tires can feel rough. A candle can be waxy. So can a crayon.

A duck has a soft, feathery body, scaly legs, and a hard bill.

More Than One

Objects can have more than one texture. Shirts can be silky. The buttons can feel bumpy. Pencils can have a pointy end and a rubbery end.

Fast Fact

Animals have many textures.

Activity
Double Trouble!

MATERIALS

Two bags
Small objects with different textures. Get two of each. For example, two marbles, two rubber balls, two cotton balls, two round erasers, two round sponges.

Can you match objects by touch? Let's get started!

Procedure:

Step 1: Place one of each object in a bag.

Step 2: Put the matching objects in the second bag.

Step 3: Pull an object from a bag.

Step 4: Use your fingers to find the matching object in the other bag.

Don't look!

This is a fun game to play with a friend. Who can find the matching object the fastest?

Objects can have a similar shape but very different textures.

Learn More

Books

Lindeen, Mary. *Texture*. Chicago, IL: Norwood House, 2018.

O'Hara, Nicholas. *Sort It by Texture*. New York, NY: Gareth Stevens, 2016.

Rompella, Natalie. *Experiments in Material and Matter with Toys and Everyday Stuff*. North Mankato, MN: Capstone, 2016.

Websites

Happy Lion Learning
www.youtube.com/watch?v=SgjNbDyIfsQ
Discover textures in this colorful video.

PBS Learning Media: What Is Texture?
www.pbslearningmedia.org/resource/npt-artquest-203-whatistexture/artquest-what-is-texture
Explore textures in a simple art lesson.

Index

activity, 22
animals, 21
atoms, 5, 7
brain, 13
chemical properties, 9
feel, 9, 11, 13
forms of matter, 7
gases, 7
liquids, 7, 11, 17
matter, 5, 7, 9
physical properties, 9, 11
skin, 13
solids, 7, 11, 15
texture, 11, 13, 15, 17, 19, 21
touch, 13

Published in 2020 by Enslow Publishing, LLC.
101 W. 23rd Street, Suite 240, New York, NY 10011

Copyright © 2020 by Enslow Publishing, LLC.

All rights reserved.

No part of this book may be reproduced by any means without the written permission of the publisher.

Library of Congress Cataloging-in-Publication Data

Names: Rector, Rebecca Kraft, author.
Title: Texture / Rebecca Kraft Rector.
Description: New York : Enslow Publishing, [2020] | Series: Let's learn about matter | Includes bibliographical references and index. | Audience: K to grade 4.
Identifiers: LCCN 2018048401| ISBN 9781978507555 (library bound) | ISBN 9781978509160 (pbk.) | ISBN 9781978509177 (6 pack)
Subjects: LCSH: Matter—Properties—Juvenile literature. | Materials—Texture—Juvenile literature. | Touch—Juvenile literature.
Classification: LCC QC173.16 .R434 2020 | DDC 530—dc23

LC record available at https://lccn.loc.gov/2018048401

Printed in the United States of America

To Our Readers: We have done our best to make sure all website addresses in this book were active and appropriate when we went to press. However, the author and the publisher have no control over and assume no liability for the material available on those websites or on any websites they may link to. Any comments or suggestions can be sent by e-mail to customerservice@enslow.com.

Photos Credits: Cover, p. 1 Captured by bondart/Shutterstock.com; p. 4 Pedro Silmon/ArcaidImages/Getty Images; p. 6 Dominique de La Croix/Shutterstock.com; p. 8 Heidi Besen/Shutterstock.com; p. 10 Sergey Orlov/Shutterstock.com; p. 12 Daisy Liang/Shutterstock.com; p. 14 Lightspring/Shutterstock.com; p. 16 mangostock/Shutterstock.com; p. 18 Alinute Silzeviciute/Alamy Stock Photo; p. 20 sh.el.photo/Shutterstock.com; p. 23 Daniel W. Slocum/Shutterstock.com; interior design elements (crumpled paper) RLRRLRLL/Shutterstock.com, (molecules) 123dartist/Shutterstock.com.